Saturn

Uranus

Neptune

# Jupiter, Ceres, and the Asteroids

Editor in chief: Paul A. Kobasa
Supplementary Publications: Jeff De La Rosa, Lisa Kwon,
    Christine Sullivan, Scott Thomas, Marty Zwikel
Research: Mike Barr, Cheryl Graham, Jacqueline Jasek,
    Barbara Lightner, Andy Roberts, Loranne Shields
Graphics and Design: Kathy Creech, Sandra Dyrlund,
    Charlene Epple, Tom Evans, Brenda Tropinski
Permissions: Janet Peterson
Indexing: David Pofelski
Pre-Press and Manufacturing: Carma Fazio, Anne Fritzinger,
    Steven Hueppchen, Tina Ramirez
Writer: Dan Blunk

First edition published 2006. Second edition published 2007.

WORLD BOOK and the GLOBE DEVICE are registered
trademarks or trademarks of World Book, Inc.

World Book, Inc.
233 N. Michigan Avenue
Chicago, IL 60601
U.S.A.

**Library of Congress Cataloging-in-Publication Data**
Jupiter, Ceres and the asteroids. -- 2nd ed.
     p. cm. -- (World Book's solar system & space exploration
library)
   Rev. ed of: Jupiter. Chicago : World Book, c2006.
   Summary: "Introduction to Jupiter, Ceres, and the Asteroids,
providing to primary and intermediate grade students
information on their features and exploration. Includes fun
facts, glossary, resource list and index"--Provided by publisher.
   Includes bibliographical references and index.
   ISBN-13: 978-0-7166-9515-8
   ISBN-10: 0-7166-9515-4
   1. Jupiter (Planet)--Juvenile literature.  2. Ceres (Planet)--
Juvenile literature.  3. Asteroids--Juvenile literature.  I. World
Book, Inc.  II. Jupiter.
   QB661.J856 2007
   523.45--dc22
                                                2006030041

ISBN-13 (set): 978-0-7166-9511-0
ISBN-10 (set): 0-7166-9511-1

Printed in the United States of America

1 2 3 4 5 6 7 8 09 08 07 06

**For information about other World Book publications,
visit our Web site at http://www.worldbook.com
or call 1-800-WORLDBK (967-5325).**

**For information about sales to schools and libraries,
call 1-800-975-3250 (United States);
1-800-837-5365 (Canada).**

**Picture Acknowledgments:** Front & Back Cover: ESA/Dave Hardy; NASA/JPL; NASA/JPL/University of Arizona;
NASA/University of Arizona/LPL; Inside Back Cover: © John Gleason, Celestial Images.

Musée du Louvre, Paris (Dagli Orti, The Art Archive) 35; © Calvin J. Hamilton 13; © Michael Carroll 29; NASA
33, 37; NASA/California Institute of Technology 23; NASA/Johns Hopkins University Applied Physics Laboratory
59; NASA/JPL 15, 17, 47, 53; NASA/JPL/Cornell University 31; NASA/JPL/Space Science Institute 39;
NASA/JPL/University of Arizona 3, 11; NASA/Lunar and Planetary Institute 49; NASA/University of Arizona/LPL 25;
© John Chumack, Photo Researchers 27; © John R. Foster, Photo Researchers 61; © Mark Garlick, Photo
Researchers 41; © SPL/Photo Researchers 45; © Detlev Van Ravenswaay, SPL/Photo Researchers 51; Deborah
Lee Soltesz, U.S. Geological Survey 57.

**Illustrations:** Inside Front Cover: WORLD BOOK illustration by Steve Karp; WORLD BOOK illustration by Steve
Karp 55; WORLD BOOK illustrations by Paul Perreault 1, 9, 21, 43; WORLD BOOK illustration by Precision
Graphics 7.

Astronomers use different kinds of photos to learn about objects in space—such as planets. Many photos show an
object's natural color. Other photos use false colors. Some false-color images show types of light the human eye
cannot normally see. Others have colors that were changed to highlight important features. When appropriate, the
captions in this book state whether a photo uses natural or false color.

# WORLD BOOK'S

## SOLAR SYSTEM & SPACE EXPLORATION LIBRARY

# Jupiter, Ceres, and the Asteroids

## SECOND EDITION

**World Book, Inc.**
a Scott Fetzer company
Chicago

# Contents

## JUPITER

# CERES

# ASTEROIDS

If a word is printed in **bold letters that look like this,** that word's meaning is given in the glossary on page 63.

# — Where Is Jupiter? —

Jupiter is the fifth **planet** from the sun in our **solar system.** Jupiter is, on average, about 484 million miles (779 million kilometers) from the sun. Jupiter is about five times farther away from the sun than Earth is.

Jupiter is the innermost of what **astronomers** call the outer planets. The other outer planets are Saturn, Uranus (*YUR uh nuhs* or *yu RAY nuhs*), and Neptune.

Jupiter's **orbit** is between the orbits of the planets Mars and Saturn. Jupiter's orbit is closest to that of Mars. But, Jupiter's orbit is, on average, about 342 million miles (550 million kilometers) farther from the sun than the orbit of Mars is.

The orbit of Jupiter's outer neighbor, Saturn, is about 407 million miles (655 million kilometers) farther from the sun than Jupiter's orbit is.

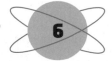

# Planet Locator

Note: The size of the sun and planets and the distance between planets in this diagram are not to scale.

Neptune

Uranus

Saturn

Jupiter

Mars

Earth

Venus

Mercury

Sun

Jupiter's symbol (top left) and a diagram showing the planet's location in the solar system

# How Big Is Jupiter?

Jupiter is massive. In fact, Jupiter is the largest **planet** in the **solar system.** Jupiter's **diameter** at its **equator** is about 88,846 miles (142,984 kilometers)—that is about 11 times the diameter of Earth. Jupiter takes up so much space that if it were hollow, more than 1,300 Earths would be needed to fill it up.

Jupiter is not only the largest of the planets in the solar system, but it also has the most **mass.** Mass is the amount of matter, or material, a thing contains. Jupiter has more mass than all of the other planets of the solar system combined.

Jupiter resembles the sun—the star at the center of our solar system—more than does any planet in the solar system. In fact, scientists think that if Jupiter had been 100 times more massive, it would have been a star.

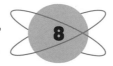

An artist's drawing comparing the size of Jupiter and Earth

Jupiter's diameter
88,846 miles
(142,984 kilometers)

Earth's diameter
7,926 miles
(12,756 kilometers)

# What Does Jupiter Look Like?

From Earth, Jupiter is one of the brightest objects in the night sky. The only objects that are usually brighter than Jupiter are Earth's **moon** and the **planet** Venus. Up close, Jupiter is covered with swirling shapes that scientists now know are actually clouds of gas. These colorful clouds, which are pushed by winds, are Jupiter's most visible feature. The gas clouds form bands of colors, ranging from orange-brown to bluish white.

Looking even closer at Jupiter, a gigantic, reddish oval object that scientists call the Great Red Spot is visible on the southern half of the planet. Scientists think the Great Red Spot, which is larger around than Earth, is a huge, swirling storm that has existed for at least 175 years, and probably for much longer.

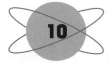

Jupiter in a
natural-color
photomosaic

# What Makes Up Jupiter?

Scientists call **planets** like Jupiter **gas giants** because they are made up mostly of gas and liquid, unlike Earth, which mainly is formed of rocky material. Jupiter consists mainly of the gases **hydrogen** and **helium.** It also contains small amounts of other substances that are made up of **oxygen, carbon, nitrogen, sulfur,** and other chemical **elements.**

Jupiter's outer layer consists chiefly of hydrogen and helium gas. Below this gas, Jupiter has a layer of liquid hydrogen and helium. This layer is present because the weight of the huge gas layer presses down on the planet, causing **atoms** to squeeze together and form into a liquid. Further down— about 6,000 miles (10,000 kilometers) below the clouds—tremendous pressure causes the liquid hydrogen to turn into an unusual form of hydrogen called liquid metallic hydrogen.

Below the liquid metallic hydrogen, scientists think Jupiter may have a dense **core** made up of substances normally found in rock and ice. This core alone is about 10 to 15 times more massive than the entire planet Earth.

# The Interior of Jupiter

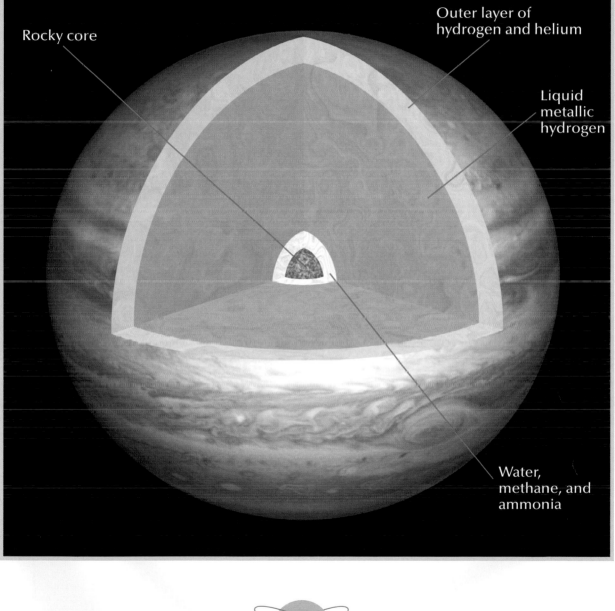

Rocky core

Outer layer of
hydrogen and helium

Liquid
metallic
hydrogen

Water,
methane, and
ammonia

13

# What Is Jupiter's Atmosphere Like?

Although Jupiter's **atmosphere** is made up mostly of **hydrogen** and **helium** "air," it is the clouds in the lower atmosphere that give the **planet** its colorful, banded appearance. The clouds that are most visible from space are made up mainly of **ammonia** (*uh MOHN yuh* or *uh MOH nee uh*) crystals.

By themselves, ammonia clouds are colorless. But, small amounts of other chemicals give Jupiter's clouds their color. In addition, slight differences in the altitude (height) of these clouds cause different bands to vary in brightness. Bands with high altitudes reflect more sunlight, appearing brighter than lower-altitude clouds.

Scientists think there may be other types of clouds beneath those made of ammonia. A second layer may contain clouds of **ammonium hydrosulfide.** Below that, there may be another layer of clouds made up of water ice.

A thunderstorm on Jupiter in a false-color photo

The white area represents a thundercloud in Jupiter's atmosphere. The upper level of the cloud was made up of ammonia crystals.

# What Is the Weather on Jupiter?

The weather on Jupiter is very cloudy and windy. Storms on the planet are famous for lasting a long time. The largest of Jupiter's storms is the Great Red Spot, a gigantic, swirling storm that has been visible from Earth since at least 1831.

The Great Red Spot measures about 7,450 miles (12,000 kilometers) from north to south and about 10,500 miles (17,000 kilometers) from east to west. The entire Earth could fit inside of this huge storm.

The winds on Jupiter blow very hard and can reach speeds of about 400 miles (650 kilometers) per hour near the **equator.** The temperature on Jupiter varies widely. The hottest region of Jupiter is deep within the **core** of the planet. In that core, the temperature can reach 45,000 °F (25,000 °C ). The temperature gets cooler and cooler toward the planet's exterior. Near Jupiter's cloud tops, the temperature is roughly -236 °F (-149 °C).

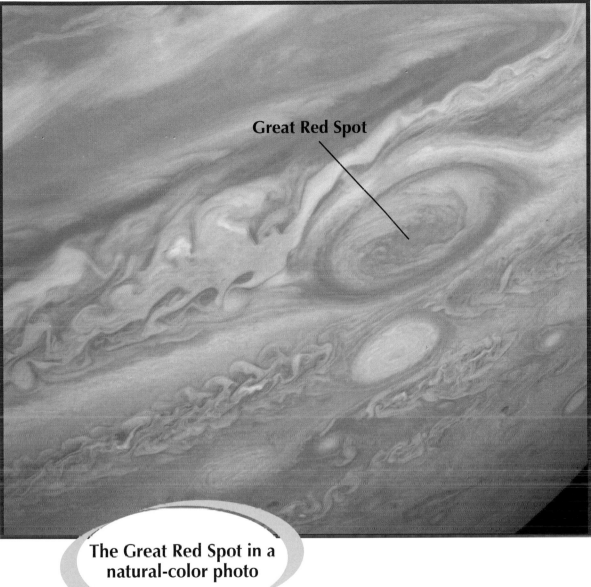

Great Red Spot

The Great Red Spot in a
natural-color photo

# How Does Jupiter Compare with Earth?

Although Jupiter has about 318 times more **mass** than Earth, its average **density** is much lower. Jupiter is only slightly more dense than liquid water on Earth.

Earth is about four times as dense as Jupiter. Jupiter's density is so much lower because it is made mostly of the light **elements**—**hydrogen** and **helium** gases. Earth is a rocky **planet** made of mostly heavier materials.

Because Jupiter has much more mass than Earth, its **gravity** is much stronger. Jupiter's gravity is about 2.4 times stronger than that of Earth. If you weighed 100 pounds on Earth, you would weigh 240 pounds on Jupiter.

# How Do They Compare?

| | Earth ⊕ | Jupiter ♃ |
|---|---|---|
| Size in diameter (at equator) | 7,926 miles (12,756 kilometers) | 88,846 miles (142,984 kilometers) |
| Average distance from sun | About 93 million miles (150 million kilometers) | About 484 million miles (779 million kilometers) |
| Length of year (in Earth days) | 365.256 | 4,332.5 |
| Length of day (in Earth time) | 24 hours | 9 hours 56 minutes |
| What an object would weigh... | If it weighed 100 pounds on Earth... | ...it would weigh about 240 pounds on Jupiter. |
| Number of moons | 1 | 16 large moons, about 50 smaller moons |
| Rings? | No | Yes |
| Atmosphere | Nitrogen, oxygen, argon | Hydrogen, helium, methane, ammonia, carbon monoxide, ethane, acetylene, phosphine, water vapor |

# What Are the Orbit and Rotation of Jupiter Like?

Jupiter takes a long time to complete one **orbit.** Jupiter is about 484 million miles (779 million kilometers) away from the sun, more than five times farther from the sun than Earth is. Because it is so far away, Jupiter takes almost 12 Earth years to make a complete orbit around the sun. That is the length of a **year** on Jupiter.

Jupiter is a fast-spinning **planet.** It rotates (spins around) faster than any other planet in the **solar system.** Earth makes a complete rotation about every 24 hours. Jupiter turns so fast that it makes a complete rotation in slightly less than 10 hours. This makes a **day** on Jupiter much shorter than a day on Earth.

Because Jupiter rotates so fast, the planet's shape is not perfectly round. Instead, it bulges slightly along the **equator** of the planet.

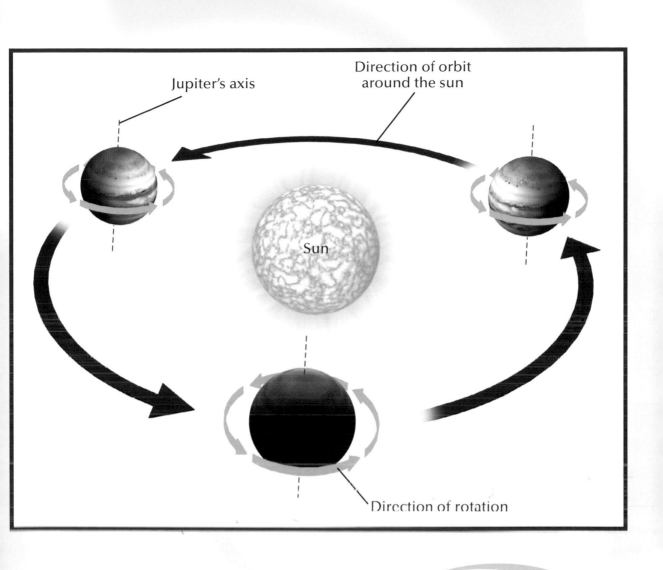

Jupiter's axis

Direction of orbit
around the sun

Sun

Direction of rotation

A diagram showing the
orbit and rotation of
Jupiter

# What Is the Surface of Jupiter Like?

To an observer from space, Jupiter's cloud tops may resemble a solid surface. But no one could stand on the surface of Jupiter. Although Jupiter has a very thick **atmosphere,** the clouds and gases are not thick enough ever to support people standing on top of them.

If Jupiter did have a surface, that surface would be a very cold and windy place. Jupiter is much farther from the sun than Earth is and, therefore, does not receive nearly as much heat from the sun as Earth does. As a result, the "surface" on Jupiter might be about 160 Fahrenheit degrees (89 Celsius degrees) colder than on Earth.

Although the high temperatures deep within Jupiter cause the gases to warm and rise upward, these gases gradually cool as they rise. The movement of these gases do, however, cause very strong winds and windstorms. Some of these windstorms can rage for hundreds of years.

A 3-D image shows the cloudy "surface" of Jupiter in false color

# How Many Moons Does Jupiter Have?

Jupiter has 16 **moons** that are at least 6 miles (10 kilometers) across. In addition to these large moons, Jupiter has dozens of smaller moons, about 50 at last count! And scientists keep discovering more moons around Jupiter.

Jupiter's moons vary widely. Some of them have an **atmosphere,** while others do not. The moons also vary in size, color, and **density.** Because of the large number of moons and their variety, **astronomers** sometimes say Jupiter and its moons resemble a mini-**solar system.**

Jupiter's moon Io in a false-color photomosaic

# What Are the "Galilean Moons"?

Four of Jupiter's **moons** are so large that they are visible from Earth with crude telescopes. In 1610, the Italian **astronomer** Galileo (*GAL uh LAY oh* or *GAL uh LEE oh*) was looking at Jupiter through a telescope and discovered the moons, later named Io *(EYE oh),* Europa *(yu ROH puh),* Ganymede *(GAN uh meed),* and Callisto *(kuh LIHS toh).*

Io, the closest of the Galilean moons to Jupiter, has the most volcanic activity of any body in the **solar system.** This moon is 2,264 miles (3,643 kilometers) in **diameter.**

Europa is an icy moon with cracks on its surface. Europa is 1,940 miles (3,122 kilometers) in diameter.

Ganymede is the largest of Jupiter's moons and, in fact, is the largest moon in the solar system. Ganymede is 3,270 miles (5,262 kilometers) in diameter.

Of the Galilean moons, Callisto is the farthest away from Jupiter. Callisto is more completely covered in **craters** than almost any object in the solar system. This moon is 2,996 miles (4,821 kilometers) in diameter.

Jupiter surrounded by the Galilean moons, in a photo taken through a telescope

Jupiter

# What Are Some of the Other Moons?

In addition to the Galilean **moons,** scientists have grouped Jupiter's other moons into two other groups: the inner **satellites** and the outer satellites. The inner satellites are closer to Jupiter than the Galilean moons are, and the outer satellites are farther away.

Most of Jupiter's moons are named after figures in Greek and Roman mythology. Here are the names of Jupiter's four inner moons:

- Metis *(MEE this),* named after the Greek goddess of wisdom.

- Adrastea *(uh DRAS tee uh),* named after the Roman distributor of rewards and punishments.

- Amalthea *(am uhl THEE uh),* named after a nymph who nursed the Greek god Zeus in his youth. Zeus is the Greek god who ruled over the other Greek gods. Jupiter is the Roman version of Zeus.

- Thebe *(THEE bee),* named after a daughter of Zeus.

**Astronomers** continue to discover additional moons of Jupiter. Some of these moons are extremely small, less than 1.2 miles (2 kilometers) in **diameter.**

An artist's conception of the surface of Amalthea, with Jupiter in the background

# Does Jupiter Have Rings?

**Astronomers** were not sure that Jupiter had rings until 1979, when the United States National Aeronautics and Space Administration (NASA) sent two space **probes**— called Voyager 1 and Voyager 2—to explore the outer planets and beyond. During their **fly-by,** the Voyager probes took pictures of what looked like two or maybe three rings.

After studying additional pictures taken with NASA's Galileo probe during the 1990's, scientists found that Jupiter has four faint rings. The brightest of these is called the main ring, a fainter ring is called the halo ring, and the two faintest rings are called the gossamer rings. All of these rings are much fainter than the rings around the planet Saturn.

Astronomers think that the rings are made of extremely tiny dust particles (pieces) that may have been knocked loose from Jupiter's inner moons by collisions with small **meteoroids.**

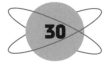

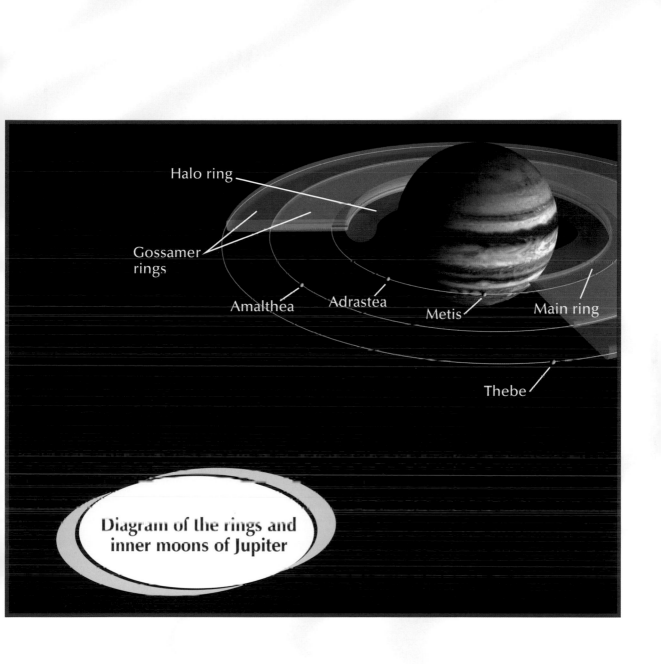

Halo ring

Gossamer rings

Amalthea

Adrastea

Metis

Main ring

Thebe

Diagram of the rings and inner moons of Jupiter

# How Fast Do Winds Blow on Jupiter?

The winds on Jupiter are very fast. Jupiter's winds are fastest along the **equator** of the **planet.** Winds on this part of the planet can reach speeds of up to 400 miles (650 kilometers) per hour.

Scientists think **convection currents** generate Jupiter's winds. This process on Jupiter can be seen from space. The pale bands of clouds on Jupiter are actually areas where the temperature is warmer, and the clouds and gases are rising. The darker bands are cooler areas where the clouds and gases are sinking.

# How Did Jupiter Get Its Name?

Ancient **astronomers** knew about Jupiter because it can be easily seen with the naked eye. These astronomers may not have known exactly what Jupiter was, but they tracked it as it moved across the night sky.

Because the **planet** was so large and bright, ancient astronomers named it after the most powerful Roman god, Jupiter. Jupiter was the god of the sky and of thunder and lightning in Roman religion. He used a thunderbolt as a weapon, and he had the power to send Earth clear weather, rain, or destructive storms. Jupiter was also worshiped as the ruler of the gods and of the universe. Jupiter is the Roman equivalent of the Greek god Zeus.

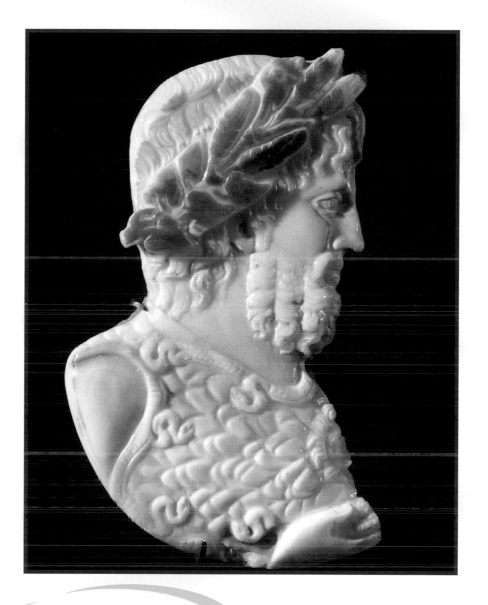

A relief sculpture of the
Roman god Jupiter

# What Space Missions Have Studied Jupiter?

Many space missions have been sent to study Jupiter. On December 3, 1973, the NASA **probe** Pioneer 10 became the first probe to fly past Jupiter. It came within 81,000 miles (130,000 kilometers) of Jupiter's clouds. In 1974, a second NASA probe called Pioneer 11 (later renamed Pioneer-Saturn) studied Jupiter. Both probes sent back information about the **atmosphere,** gravitational pull, and **magnetic field** of Jupiter.

In 1979, the NASA probes Voyager 1 and Voyager 2 flew past Jupiter. These probes discovered much of what we know about Jupiter's atmosphere.

NASA's Ulysses probe, whose main purpose was to study the sun, swung within 235,000 miles (378,000 kilometers) of Jupiter in 1992 and studied its radio emissions.

Probably the most important mission to Jupiter was NASA's Galileo probe. Galileo reached Jupiter in 1995 and became the first probe to **orbit** the **planet.** Galileo released a small probe that became the first instrument to sample the atmosphere of a **gas giant.**

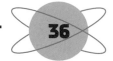

An artist's drawing of the Galileo probe with Io (left) and Jupiter in the background

# Could There Be Life on Jupiter or Its Moons?

Jupiter has no real surface because it is made mostly of gas. In addition, the **planet** has a very strong gravitational pull, extreme temperatures, and no bodies of water. Scientists do not think life as we know it could exist on Jupiter.

However, some of Jupiter's **moons** may have the conditions needed for life. For example, Europa is an icy moon that many scientists think may have an ocean of liquid water beneath its frozen surface. And, where there is water, there could be living organisms.

Europa is too far from the sun to make the surface warm enough to melt water ice. However, scientists think that the **gravity** of Jupiter and Jupiter's other large **satellites** pulling Europa in different directions could produce enough heat inside this moon to allow liquid water to exist.

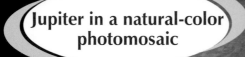

Jupiter in a natural-color photomosaic

# What Is Ceres?

An **asteroid?** A **planet?** A **dwarf planet?** Ceres *(SIHR eez)* is certainly an interesting object in the **solar system.**

Ceres is an asteroid, a small body made of rock, **carbon,** or metal that **orbits** the sun. Ceres is made of rock and water ice. The surface of Ceres most likely is made of carbon-rich rock under a layer of ice.

Ceres is the largest asteroid in the **asteroid belt.** Ceres is so large that when it was first discovered (see page 44), it was thought to be a planet. The longest **diameter** of Ceres is 596 miles (960 kilometers) and the shortest is 579 miles (932 kilometers). That makes Ceres less than ⅓ the size of Earth's **moon.**

In addition to being an asteroid, in 2006 Ceres began to be considered a dwarf planet. In that year the International Astronomical Union (IAU) adopted a new definition for what could be called a planet, and they also created a new category of objects called dwarf planets. To be a dwarf planet, the IAU decided that an object had to orbit the sun and be round. Ceres meets both qualifications. Not all **astronomers** agree with the new dwarf-planet category, however, so Ceres's classification could change again.

An artist's drawing
of Ceres

# Where Is Ceres?

Ceres is in the **asteroid belt,** between the **orbits** of Mars and Jupiter. That puts Ceres at the outer edge of the inner **planets** (Mercury, Venus, Earth, and Mars).

The orbit of Ceres is closest to that of Mars. Mars is, on average, about 142 million miles (228 million kilometers) from the sun. The average distance from the sun to Ceres is around 257 million miles (414 million kilometers). The average distance from the sun to Jupiter is around 484 million miles (779 million kilometers).

That makes the average distance between Ceres and its inner neighbor, Mars, about 115 million miles (186 million kilometers). The average distance between Ceres and its outer neighbor, Jupiter, is about 227 million miles (365 million kilometers).

In 2006, NASA planned to launch Dawn, a mission to Ceres, in 2007. According to the mission plan, Dawn will **fly by** Mars in 2009. After a visit to another **asteroid,** Vesta, Dawn will reach Ceres in 2015.

Jupiter

Ceres

Asteroid belt

Mars

An artist's drawing
showing the location
of Ceres

# Who Discovered Ceres?

An Italian **astronomer,** a monk named Giuseppe Piazzi *(joo ZEHP peh PYAHT tsee),* first spotted Ceres in 1801. He named the object for the Roman goddess of grain and the harvest. Piazzi tracked Ceres for several weeks but then lost the object. In the fall of 1801, the German mathematician Carl Friedrich Gauss *(FREE drihk GOWS)* predicted the place in the sky where astronomers should look to find Ceres again. They found Ceres at the end of 1801. Ceres was the first of the many objects discovered that would eventually be called **asteroids**.

In 1802, the British astronomer William Herschel *(HUR shuhl)* introduced the word *asteroid* to apply to Ceres and another object found in that year, Pallas. But, most astronomers considered Ceres, Pallas, and similar objects discovered later, to be **planets.** Only when many of these small objects had been found did astronomers begin to adopt the concept of "asteroid" to describe them. Ceres lost its status as a planet and became an asteroid.

In 2006, the status of Ceres changed again, as it became classed as both an asteroid and a **dwarf planet** (see page 40).

An illustration of
Giuseppe Piazzi

# What Are Jupiter's Rocky Neighbors?

Jupiter's rocky neighbors are **asteroids!** Asteroids are irregularly shaped objects that were left over from the formation of the **solar system** billions of years ago. Scientists estimate that there are millions of asteroids.

Most asteroids are made of metals or rocky material or are rich in **minerals** containing **carbon.** Like **planets,** asteroids rotate as they **orbit** the sun. But asteroids do not have many of the characteristics of planets, such as an **atmosphere.**

Asteroids can range in size from nearly 600 miles (965 kilometers) in **diameter** to tiny asteroids less than 20 feet (6 meters) across. Some asteroids are large enough to have their own **moon.** For example, the asteroid Ida has a small moon named Dactyl.

Dactyl

Ida

**Ida and its moon, Dactyl**

# What Is the Asteroid Belt?

Most of the **asteroids** in our **solar system** are found in one area—between the **orbits** of the **planets** Jupiter and Mars. This region is known as the **asteroid belt,** or the Main Belt.

Scientists have discovered that there are two parts of the Main Belt and that different types of asteroids are in each area. The outer part of the Main Belt contains asteroids that are rich in **carbon**—a chemical **element.** These asteroids are very old and have not changed much since the solar system formed.

The asteroids in the inner part of the Main Belt, which is closer to Earth, contain many metal-rich **minerals.** Scientists think these asteroids were formed in very high temperatures.

Located outside the Main Belt and closer to Jupiter are more than 1,000 other asteroids. These are called Trojan asteroids. They are named for the heroes of the Trojan War in Greek legend.

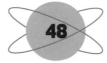

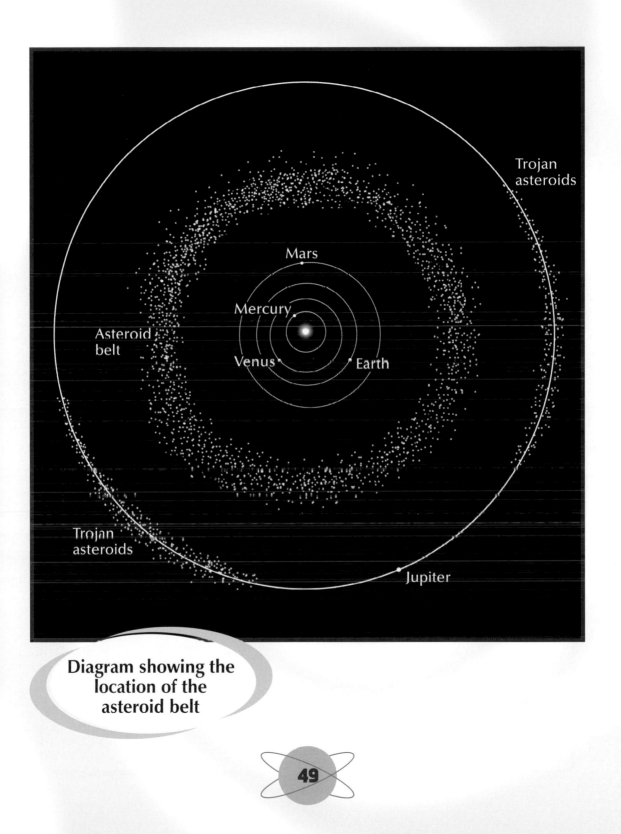

Trojan
asteroids

Mars

Mercury

Asteroid
belt

Venus        Earth

Trojan
asteroids

Jupiter

**Diagram showing the
location of the
asteroid belt**

# What Is the Connection Between Jupiter and the Asteroid Belt?

The **asteroid belt,** or Main Belt, exists near Jupiter because Jupiter has so much **mass.**

In the late 1700's, **astronomers** expected to find a **planet** in the area between the **orbits** of Jupiter and Mars. At that time, the orbits of all the known planets in the **solar system** followed a mathematical pattern. (Later discoveries, such as Neptune and the **dwarf planet** Pluto, would break that pattern, somewhat.) Instead of finding a planet between Jupiter and Mars, however, astronomers began to find **asteroids.**

Asteroids are made of the same material that formed the planets. Most scientists now think that the rocky matter in the asteroid belt might eventually have formed into a planet, but the amazingly strong force of Jupiter's **gravity** prevented it. And Jupiter's gravity now keeps many of the asteroids in the outer asteroid belt from wandering out of the region.

An artist's drawing of the solar system, showing the asteroid belt between the orbits of Mars and Jupiter

# What Are the Surfaces of Asteroids Like?

**Asteroids** have many different kinds of surfaces. Some are dark. Others are very bright because they reflect much of the light they receive from the sun. The way asteroids look has a lot to do with the materials of which they are made.

For example, dark asteroids are often made of substances rich in **carbon.** Brighter asteroids contain metal-rich **minerals** that reflect the sun's light. This shiny surface makes them more visible to scientists studying asteroids.

Some asteroids even have "mountains" on them. One such asteroid is named Vesta. In 1996, the Hubble Space Telescope took a picture of Vesta that showed a huge impact **crater.** Scientists think that a large object struck Vesta, creating the crater. The impact caused so much heat that a "mountain peak" was formed in the center of the crater as molten (melted) material flowed back into the crater.

The asteroid Gaspra in a black-and-white photomosaic

# Are All Asteroids Found in the Asteroid Belt?

Although the vast majority of **asteroids** are near Jupiter in the **asteroid belt,** or Main Belt, some are orbiting in other regions of the **solar system.** Many of these asteroids are known as near-Earth asteroids. These near-Earth asteroids can be grouped according to their **orbit.**

- The Atens are asteroids that have orbits that lie mostly or totally inside the orbit of Earth.

- The Apollos are asteroids with orbits that occasionally cross Earth's orbit.

- The Amors are asteroids with orbits that cross the orbit of Mars but not of Earth.

Each of these types of asteroids is named for a single asteroid of the same name that is typical of the type. That is, the Atens are named for a specific asteroid named Aten.

There are also asteroids called Trojans (see the diagram on page 49) that follow the same orbit as Jupiter. And there are a few asteroids in the outer areas of the solar system. These are called Centaurs, and at least some of them could be **comets** instead of asteroids.

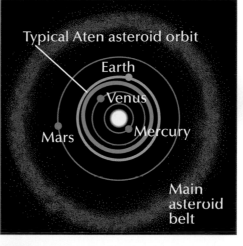

Typical Aten asteroid orbit

Earth

Venus

Mars

Mercury

Main asteroid belt

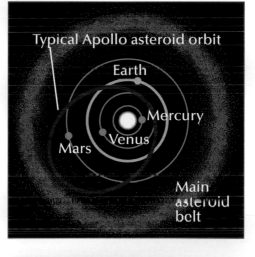

Typical Apollo asteroid orbit

Earth

Mercury

Mars

Venus

Main asteroid belt

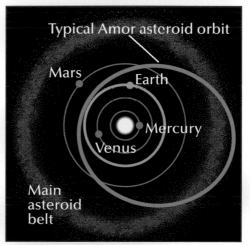

Typical Amor asteroid orbit

Mars

Earth

Mercury

Venus

Main asteroid belt

Typical orbits for three types of near-Earth asteroids

# What Can We Learn from Asteroids?

Scientists have learned quite a bit about the history of our **solar system** from **asteroids.** And most of what we know about asteroids comes from studying **meteorites.**

Meteorites are masses of stone or metal that have reached Earth from outer space without burning up. Most scientists agree that the majority of meteorites are fragments that splintered off from asteroids.

Asteroids are very interesting to scientists. Asteroids are very old, and most of them have not changed in billions of years. For these reasons, asteroids could tell scientists a great deal about how the solar system formed. For example, by looking at what asteroids are made of, scientists can learn what types of materials were common when the solar system was first formed.

Meteor Crater, in Arizona, was formed when a meteorite struck Earth

# What Space Missions Have Been Sent to Study Asteroids?

Before 1991, the only way scientists could study **asteroids** was by using telescopes from Earth. Since then, several space missions have sent **probes** to study asteroids in our **solar system.**

In 1991, the NASA space probe Galileo took the first close-up pictures of an asteroid, Gaspra. Galileo went on to study the asteroid Ida in 1993 and detected its moon, Dactyl.

In 1997, NASA's Near-Earth Asteroid Rendezvous *(RAHN duh voo),* or NEAR, probe studied the asteroid Mathilda and discovered numerous deep impact **craters.** In February 2000, the NEAR probe made history by going into **orbit** around the asteroid 433 Eros. In that same year, the probe was renamed NEAR-Shoemaker, in honor of American astronomer Eugene Shoemaker (1928-1997).

On February 12, 2001, NEAR-Shoemaker, reached the surface of Eros, taking a final picture as it landed.

An artist's drawing of the NEAR-Shoemaker probe (above) and two images of Eros

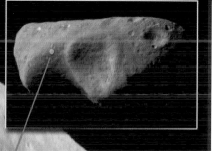

The site where NEAR-Shoemaker touched down on Eros

# Are Asteroids Dangerous?

**Asteroids** are not usually dangerous. However, there is a danger that a large asteroid could change its **orbit** around the sun and strike the Earth. This has happened in Earth's history. In fact, scientists think that an asteroid striking the Earth about 65 million years ago may have caused such destruction to Earth's environment that it caused the dinosaurs to become extinct.

In 1989, an asteroid called 1989 FC—later named Asclepius—came very close to Earth. This convinced many scientists that they should begin to track asteroids near Earth. In 1995, NASA began its Near-Earth Asteroid Tracking program. The scientists in this program look for asteroids that could collide with Earth and watch them to see if they are a danger to Earth. Scientists are even discussing plans of what to do if such an asteroid were on a course to strike Earth.

An artist's drawing of an imagined collision between an asteroid and Earth

# FUN FACTS
## About JUPITER, CERES, & the ASTEROIDS

★ Different parts of Jupiter rotate at different speeds. The material at Jupiter's **equator** rotates faster than the material at its poles.

★ The storm known as the Great Red Spot on Jupiter is shrinking.

★ In 1994, scientists were able to observe a rare event on Jupiter. As the **comet** Shoemaker-Levy 9 approached Jupiter, that planet's powerful **gravity** caused the comet to break up. Scientists were able to use a number of telescopes and **probes** to watch the pieces of the comet collide with Jupiter.

★ The time it takes for Ceres to rotate on its axis is a little more than 9 hours in Earth time. It takes Ceres about 4.5 Earth years to **orbit** the sun.

★ The **asteroid belt,** which lies just inside Jupiter's orbit, is very large. The average distance between the **asteroids** there is around 62,000 miles (100,000 kilometers). If you could travel in the asteroid belt, you would be unlikely to see an asteroid while you were there, as there is far more space than asteroid in the belt.

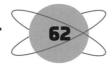

# Glossary

**ammonia** A compound made up of nitrogen and hydrogen.

**ammonium hydrosulfide** A substance made of nitrogen, hydrogen, and sulfur.

**asteroid** A small body made of rock, carbon, or metal that orbits the sun. Most asteroids are between the orbits of Mars and Jupiter.

**asteroid belt** The region between Jupiter and Mars where the majority of the asteroids exist.

**astronomer** A scientist who studies stars and planets.

**atmosphere** The mass of gases that surrounds a planet.

**atom** One of the basic units of matter. Atoms are incredibly tiny—more than a million times smaller than the thickness of a human hair.

**carbon** A nonmetallic chemical element.

**comet** A small body made of dirt and ice that orbits the sun.

**convection current** Movement or circulation that happens—for instance, in a planet's atmosphere—when warm gases rise and cooler gases sink.

**core** The center part of the inside of a planet.

**crater** A bowl-shaped depression on the surface of a moon or planet.

**day** The time it takes a planet to rotate (spin) once around its axis and come back to the same position in relation to the sun.

**density** The amount of matter in a given space.

**diameter** The distance of a straight line through the middle of a circle or anything shaped like a ball.

**dwarf planet** A round body in space orbiting a star, which does not have enough gravitational pull to clear other objects from its orbit.

**element** A chemical element is any substance that contains only one kind of atom.

**equator** An imaginary circle around the middle of a planet.

**fly-by** Flight whereby a spacecraft flies near to an object in space but does not land on or orbit that object.

**gas giant** Any of four planets—Jupiter, Saturn, Uranus, and Neptune—made up mostly of gas and liquid.

**gravity** The effect of a force of attraction that acts between all objects because of their mass (that is, the amount of matter the objects have).

**helium** The second most abundant chemical element in the universe.

**hydrogen** The most abundant chemical element in the universe.

**magnetic field** The space around a magnet or magnetized object, within which its power of attraction works.

**mass** The amount of matter a thing contains.

**meteorite** A mass of stone or metal from outer space that has reached the surface of a planet without burning up in that planet's atmosphere.

**meteoroid** A small object, believed often to be the remains of a disintegrated comet, that travels through space.

**mineral** An inorganic (nonliving) substance made up of crystals.

**moon** A smaller body that orbits a planet.

**nitrogen** A nonmetallic chemical element.

**orbit** The path that a smaller body takes around a larger body, for instance, the path that a planet takes around the sun. Also, to travel in an orbit.

**oxygen** A nonmetallic chemical element.

**planet** A large, round body in space that orbits a star. A planet must have sufficient gravitational pull to clear other objects from the area of its orbit.

**probe** An unpiloted device sent to explore space. Most probes send data (information) from space.

**satellite** A natural object that orbits a planet or asteroid.

**silicate** A group of minerals that contain silicon, oxygen, and one or more metallic elements.

**solar system** A group of bodies in space made up of a star and the planets and other objects orbiting around that star.

**sulfur** A nonmetallic chemical element.

**year** The time it takes a planet to complete one orbit around the sun.

# Index

**For more information about Jupiter, Ceres, and the asteroids, try these resources:**

**Asteroids:**
*Asteroids, Comets, and Meteors*, by Ron Miller,
    21st Century, 2004

**Jupiter:**
*Destination: Jupiter*, by Seymour Simon, Sagebrush, 2001

*The Far Planets*, by Robin Kerrod, Raintree, 2002

*A Look at Jupiter,* by Ray Spangenburg and Kit Moser,
    Franklin Watts, 2002

**Asteroids:**
http://solarsystem.nasa.gov/planets/profile.cfm?Object
    =Asteroids&Display=Overview

**Ceres:**
http://dawn.jpl.nasa.gov/

**Jupiter:**
http://pds.jpl.nasa.gov/planets/choices/jupiter1.htm

http://www2.jpl.nasa.gov/sl9/

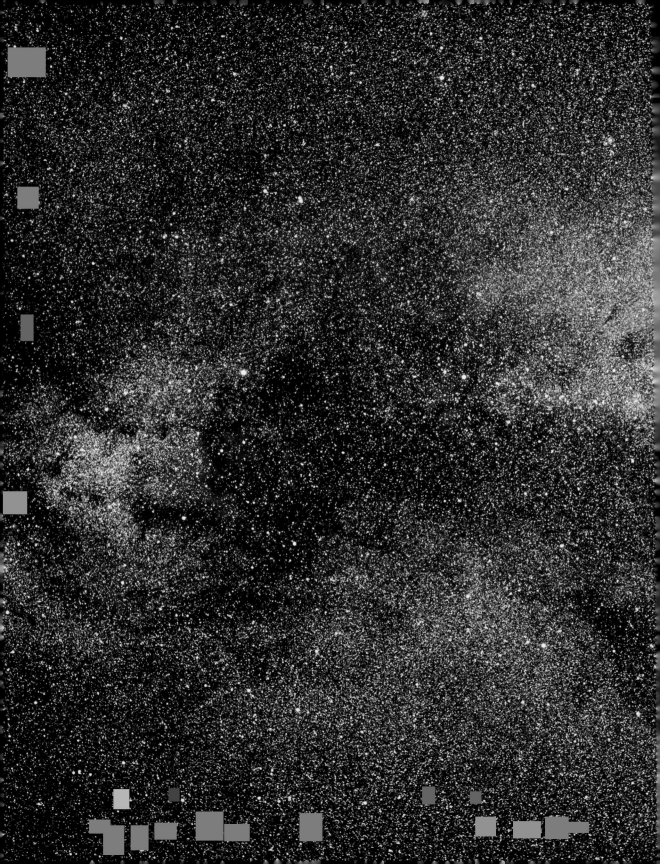